The Gardens of

MAINAU FLOWER ISLAND

The Gardens of

MAINAU FLOWER ISLAND

Text: Markus Zeiler
Photographs: Peter Allgaier

Ulmer

INTRODUCTION

Mainau Island with its nickname "Flower Island in Lake Constance" is famous far beyond the German border. With its great diversity of exceptional plants and an average of 1.2 million visitors every year, the Botanic Garden and castle grounds are among the most significant and most frequently visited gardens in Germany. The island is a paradise for plant and garden lovers as well as a place to relax for both tourists and locals.

The numerous Protected Area Categories aiming to safeguard the precious nature and cultural landscape show that Mainau and its surroundings are an absolute jewel and of great importance in many respects: The island is listed in the heritage register as a whole, in parts and also as a cultural monument of particular importance. Furthermore, the whole island is situated inside the landscape conservation area "Bodenseeufer" (Lake Constance shore) and where the bridge links the island to the mainland, Mainau is bordered by two nature reserves serving as an important retreat for birds.

From a geological point of view, Mainau Island consists of molasse sandstone. During the different ice ages which formed Lake Constance, the rock that is today Mainau Island remained untouched by the destructive force of the ice and was covered with gravel and sand. Today, the island is approx. 395 metres above sea level on its shoreline and at the highest point of the park it is up to 425 metres above sea level.

LAKE CONSTANCE AS AN IMPORTANT CLIMATE PARAMETER

With its large amount of water, Lake Constance has a great impact on Mainau's climate. In spring, when the water of this natural air conditioning system is still cold and only warming up slowly, the plants have a slower bud growth compared to other, similar regions, because the air temperature is also slower to warm up. This reduces the risk of late frost which is particularly beneficial for cold-sensitive plants. In summer, the sun heats up Lake Constance and in the autumn, this energy warms up the surrounding air as temperatures fall. As the famous Lake Constance fog forms, a number of trees and shrubs growing on climate borders now have the possibility to let their wood mature in preparation for the upcoming winter.

Climate data collected in the reference period from 1981 to 2010 show an average temperature of 9.4° Celsius and annual rainfall amounting to 830 mm. Even if Mainau Island is well-known for its subtropical vegetation in some parts, the island still sees low temperatures, ice and snow in winter. Temperatures can even drop to lows in the negative double-digits as was last the case in February 2012 over a period of several days. Usually the snow is very wet and heavy which is a problem, especially for evergreens.

Mainau Island and the area around it was settled relatively early. There are signs of pre-historic stilt houses in the shallow water. Since the beginning of the 8th century, the is-

Surrounded by exotic summer plants like *Ricinus communis* ('Carmencita Bright Red') and *Erythrina crista-galli*, the Grand Duke Frederick I of Baden watches over his park. After buying Mainau in 1853, he gradually turned it into a stately garden, thus laying the foundations for today's castle garden and the Botanic Garden.

land belonged to the eminent Monastic Island of Reichenau, situated about 10 kilometres to the south in the so-called Untersee. From 1272, Mainau belonged to the Order of Teutonic Knights. The Knights set up a base, a so-called commandery, which belonged to the Bailiwick (administrative unit) of Alsace-Burgundy.

The Teutonic Knights lived on Mainau Island for over 500 years and in that time agriculture was the economic base. Wine grapes were grown on the southern slopes, there were extensive fruit orchards on the plateau and grainfields and meadows near the lakeshore. Old maps show geometrically arranged gardens; their purpose was probably to feed the residents. With the dissolution of the Order of Teutonic Knights, Mainau became part of the newly formed Grand Duchy of Baden in 1806. But the grand dukes had little interest in the island so finally the Hungarian Prince Nicolas Esterházy bought it in 1827. Useless buildings near the castle were torn down as well as the fortifications which had become obsolete. He introduced the first foreign trees and shrubs to Mainau Island.

In 1830, a *Liriodendron tulipifera* was planted on the border to today's Arboretum. The fig espalier on the southern façade of the castle is also said to go back to plants Esterházy brought to Mainau.

Prince Esterházy's descendants sold the island to Countess Katharina of Langenstein in 1839, she sold it to her daughter and in 1853, the future Grand Duke Frederick I of Baden bought Mainau and used it as summer palace.

A GREEN PARADISE FOR THE GRAND DUKE OF BADEN

Frederick I of Baden who married the Prussian princess Louise in 1856, daughter of the future Emperor William I, was a passionate plant collector and dendrologist. Immediately after buying the island, he continued what Nicolas Esterházy had started by tearing down more buildings and fortifications and creating new paths and seating areas. In the process, more or less exotic trees and shrubs were gradually planted on the island. As soon as 1856, an alley of mulberry trees was created in order to breed silkworms. When their daughter Viktoria was born in 1862, Frederick I and Louise planted a large-leaved lime tree known today on Mainau Island as the Lime-Tree "Viktoria". Today's Arboretum saw its beginnings in 1864 when the first trees were planted.

Grand Duke Frederick I died in 1907, a very old man, on his beloved Mainau Island. Out of respect for her husband's work, his widow ordered that nothing shall be changed in the park that was now finished. After implementing this for 25 years, the park had become overgrown.

THE SWEDISH PRINCE

The son of Louise and Frederick I, the Grand Duke Frederick II of Baden, passed Mainau to his sister Viktoria. In 1881, Viktoria of Baden married the future King of Sweden Gustav V and since 1907 she was Queen Viktoria of Sweden. When she died in 1930, her son William inherited the island and he transferred the administration of Mainau to his only son Lennart. He married the Swedish commoner Karin Nissvandt in 1932, thus losing all titles and rights as Prince of Sweden. The couple decided to turn the island in Lake Constance into their home and to make a living from it. Only in the 1950s he received the title of Count af Wisborg by the Grand Duchess Charlotte of Luxemburg.

So they started to change the grand ducal park into an international tourist attraction. First Lennart Bernadotte – as he once wrote himself – started to clean up the swimming jungle. By removing numerous trees and shrubs which were invading the valuable exotic plants, he gave the park a new, attractive look. At the same time, he increasingly opened the island to the public. The visitors wanted to buy postcards and have a coffee on the island: the tourist attraction Mainau was born.

After the Second World War, the Botanic Garden was gradually extended. In the process, the Italian Rose Garden was enlarged, the Walk of Wild and Shrub Roses was created and the Italian Style Floral and Water Cascade was constructed.

Mainau became more and more popular and especially in the 1950s and 1960s, the so-called German Adriatic Sea was the place to go for a holiday. Count Lennart Bernadotte, who had re-established the German Horticultural Society at the request of Heinrich Luebke, transferred in the early 1980s the management of the company to his second wife Sonja whom he had married in 1972. With her he founded the Lennart Bernadotte Foundation (Lennart-Bernadotte-Stiftung), today sole shareholder of the tourism company Mainau GmbH.

In the time of Countess Sonja Bernadotte, who followed her husband at the head of the German Horticultural Society, the Butterfly

After the death of Countess Sonja Bernadotte af Wisborg in October 2008, the "Sonja and Lennart Bernadotte Square" was created by the Archway Building, informing about the life of the two Mainau designers. Embedded in a sea of Triumph Tulips ('Rem's Favourite', 'Jan Reus', 'Black Jack', 'Gavota' and 'Havran'), the two busts are facing the castle.

House was constructed in 1996 and the Palm House in 1998 among other things. Mainau Island's mainland entrance was redesigned, restaurants were refurbished and events such as a garden and lifestyle market and open-air concerts were initiated. Totally respecting her husband's tradition, Countess Sonja Bernadotte managed the company Mainau GmbH with the aim of enthusing people about nature and gardening.

When Count Lennart Bernadotte died in 2004 aged 95 and Countess Sonja Bernadotte in 2008 aged only 64, their two eldest children, Countess Bettina Bernadotte and Count Bjoern Bernadotte, took over the company Mainau GmbH. True to their parents' philosophy, they are now in charge of the Flower Island and undertake new projects for the development of Mainau Island as a place of recreation, relaxation and inspiration for gardening.

13
17
1 Castle and Church, Palm House
2 Gardener's Tower and Archway Building
3 Italian Rose Garden
4 Mediterranean Terraces and Fountain Arena
5 Arboretum
6 Comturey and Harbour
7 Shore Garden
8 Italian Style Floral and Water Cascade
9 Dahlia Garden
10 Perennial Garden
11 Walk of Wild and Shrub Roses
12 Pond
13 Lauen Area
14 Platanenweg 5 and Insect Garden
15 Butterfly House
16 Kitchen Garden
17 Garden for Everyone

N
1
2
3
4
5
6
7
8
9
10
11
12
14
15
16

AT THE CASTLE

The Baroque ensemble of Teutonic Order Castle and St. Marien church constitutes the centre of the Mainau park and the frame for the presentation of foreign and exotic plants. Johann Caspar Bagnato, master builder for the Teutonic Order, built the ensemble between 1732 and 1746.

The Archway Building opposite the castle, the so-called Gardener's Tower and the remaining structure of the castle wall characterize the central area of the castle, which used to be fortified, called the "inner garden" at the time of Grand Duke Frederick I. by the Archway Building and along the southwestern part of the former castle wall, the design of the classic flower beds contrasts with the Baroque splendour of the castle complex. Like in all other changing flower beds on the island, flowers are planted here twice a year. In October, bulbs like tulips, daffodils and crocus in many different varieties are assembled with spring bloomers such as *Bellis perennis, Viola cornuta* or *Mysotis sylvatica* to create complex flower arrangements. From the beginning of May, these areas are cleared and replanted with summer flowers.

On the big lawn in front of the castle, a geometric flower bed represents Baroque symmetry with its ornamental planting design. The three wings of the castle form a courtyard, an atmospheric place displaying the gardeners' skills and the nursery's botanical treasures such as a collection of historical citrus varieties.

Such treasures can also be found in the Palm House on the south terrace of the castle. This entirely removable orangery building contains a collection of firmly planted stately palm trees. The first building preceding today's Palm House was built as early as 1867 between the south wing of the castle and today's Italian Rose Garden.

Previous page: With a total surface of almost 45 hectares, Mainau is the smallest of the three large islands in Lake Constance, after Reichenau and Lindau. Located on the southeastern end of the so-called Ueberlinger Lake, the island is just below the municipality of Litzelstetten which was incorporated into the city of Konstanz (Constance) in 1971.

A belt of tulips (in the front 'Page Polka', 'Peppermintstick', 'Rosalie' and 'Artist') and spring flowers such as *Myosotis sylvatica* and *Viola cornuta* in matching colours points viewers in the direction of the Gardener's Tower and the rambler rose 'New Dawn'.

The so-called Flower Garden in front of the castle, a changing flower bed of just under 500 m² between the Gardener's Tower and the Italian Rose Garden, displays an arrangement of banana trees, flowering shrubs and annual summer flowers. The combination of annual summer flowers and conventional pot plants is a particularity in the flower bed design of the island's park. All these specimen plants are removed from the flower beds in autumn, planted in pots and taken to the greenhouses of the Mainau nursery to overwinter.

The Palm House on the southern terrace of the castle protects the dug up date palms and other plants from cold temperatures in winter. With its bright and generous architecture in combination with an appealing planting design, the house is an ideal place for the annual orchid show in spring. In line with the island's annual motto and a botanical exhibition with a particular focus, the gardeners show a part of the orchid collection not usually accessible to the public.

On the access road to the castle the visitor is greeted by a giant tuft of *Phyllostachys bambusoides* 'Violascens'. Planted in 1905, it has braved the weather repeatedly in the past 110 years. Especially snow constitutes a risk of culm fracture.

Since the mid-1990s, Mainau is in possession of a collection of historical citrus varieties grafted on high-stem *Citrus aurantium*. The garden architect Wolfgang Hundbiss gathered the 50 varieties from different Italian gardens, thus creating a hand-picked collection of noble varieties.

The large *Ginkgo biloba* overlooks the autumnal Italian Rose Garden. Native to China, the tree was planted in 1872 and its majestic appearance and golden yellow autumn colours contrast very nicely with the castle church St. Marien.

THE ITALIAN ROSE GARDEN

The Italian Rose Garden in its original form was created in the 1860s and initially completed in 1871 as an Italian style "Flower Garden". While on the north side the balustrade wall with its curved double staircase delimits the garden, on the west and east side arcades border the garden and on the south side a pergola marks the boundary. Covered abundantly with rambler roses, they constitute a delightful frame for the strictly geometrically designed historical part of the Italian style garden.

The garden came into existence at the time the rose variety 'La France' was introduced in France (1867), a turning point in the production of roses, moving away from old to modern roses. Therefore it is most likely that on Mainau Island there were roses from the beginning in the "Italian Style Flower Garden" and that new roses were regularly on display. Hence, it is not surprising that in 1886, Ludwig von Ompteda called it a rose garden in his book "Rhenish Gardens".

The design has not changed much since: Standard roses in flower beds planted with short-stem roses and hybrid tea roses. In the 20th century under Count Lennart Bernadotte af Wisborg, the garden was expanded twice according to plans by Josef Raff, the Mainau garden director at the time and rose lover: In 1961, the garden was extended to the south, incorporating the existing big fountain, and in 1980, the so-called novelties section was created at the western end. In this section, new rose varieties are constantly being planted – some of them were even baptized on Mainau – and displayed for a maximum of 5 years. When choosing new roses for the garden, the following three criteria are observed: high disease resistance, beautiful blooms and, if possible, good scent.

On Mainau, the month of June is entirely dedicated to the rose. The rambler roses on the balustrade mark the beginning of the main blooming season of the so-called Queen of Flowers in the Italian Rose Garden. The wall made of Rorschach Sandstone is covered in blossoms, forming a medley of colours. The varieties 'Hella' and 'Florentina' frame the marble sculpture, an allegory of spring.

From the south terrace of the castle, the visitor has an excellent view of the garden: Since its completion in 1871, the core area of the Italian Rose Garden has hardly changed. The narrow flower beds are still surrounded by lawn and display long-stem roses, hybrid tea roses, ground cover roses and small shrub roses – and standing out between them are standard roses with a stem height from 120 to 140 cm.

The hybrid tea rose 'Graefin Diana' stands out with its pleasant fragrance, beautiful flowers and healthy foliage. This novelty was baptized in 2012 and given the name of Countess Sonja Bernadotte's and Count Lennart Bernadotte's youngest daughter.

Following page: On the pergola, rambler roses are reaching up to the sky, such as the 'Laguna' (centre) and 'Jasmina' (right) varieties. At their feet, the small shrub rose 'Bad Birnbach' is spreading out.

MEDITERRANEAN TERRACES AND FOUNTAIN ARENA

Located on the southern side of the Mainau plateau, the Mediterranean Terraces with the Fountain Arena are among the sunniest spots on the Flower Island. In clear weather, the visitors can enjoy a wide view over Lake Constance and the Swiss village of Appenzell as far as the peaks of the Alps. From 1964 to 1968, the Mediterranean Terraces were created as a garden on four levels. The steps are made out of gneiss from Val Calanca in Ticino and were a personal contribution of the island's gardeners. Upon completion of the garden as a whole and on the occasion of Count Lennart Bernadotte's 70th birthday, the Fountain Arena with the so-called Swan Fountain ("Schwanenbrunnen") was completed in 1979. Ten years before, a sundial had already been set up on one of the terraces.

In this sunny spot, from early spring and well into autumn, plants like *Trachycarpus fortunei*, *Quercus turneri* 'Pseudoturneri', *Cupressus sempervirens* or *Choisya ternata* create a Mediterranean feel together with the annuals and exotic plants in the flower bed. In summer, the passion flower collection of the island's nursery is also on display – a collection comprising to this day over 50 species and varieties of this tropical plant.

In summer, the visitor has an unobstructed view from the Fountain Arena over the lake as far as the Alpine peaks on the Austria-Switzerland border. At the centre of the impressive garden is the finely wrought Swan Fountain, created by Dr. Marianne Kiesselbach. The royal blue flowers of Agapantus in front of the fountain emphasize the Mediterranean flair.

The thermophile *Campsis radicans* thrives *well* on the Mediterranean Terraces and has already developed a gnarled trunk. The sunny location and the dark gneissic rock greatly appeal to the climber native to Eastern North America.

Snow and temperatures below 0° Celsius are not unusual for Lake Constance and Mainau Island, so that exotic plants like *Trachycarpus fortunei* are occasionally covered with snow, but generally not for very long.

Following page: The dark colour of the gneiss from Ticino particularly emphasizes the bright yellow and orange tones of the spring bedding plants by the Fountain Arena. In the background, the Darwin Hybrid Tulips on the lawn by the Swede's Tower are in full bloom.

THE ARBORETUM

The Arboretum with a surface of almost five hectares and some stately trees is a paradise for tree lovers and offers peace and quiet all year round. From 1864, the Grand Duke Frederick I started to plant foreign trees in this area which had until then been agricultural land. And over the past 150 years, an interesting and considerable collection came into existence. As the Grand Duke particularly loved conifers, the tree collection mainly consists of evergreen trees with needles. The most important representatives are *Sequoiadendron giganteum, Cedrus atlantica, Cedrus atlantica* 'Glauca' and *Chamaecyparis*. Furthermore, there are some rare conifer species such as *Abies grandis* and *Abies firma*, *Cryptomeria japoncia* and varieties, *Calocedrus decurrens, Torreya californica* and *Sequoia sempervirens*. But in between all these coniferous trees, there are also just as interesting deciduous trees such as *Hovenia dulcis*, a *Fagus sylvatica* 'Asplenifolia' planted in honour of Count Lennart Bernadotte in 1969 and a *Quercus rubra* from 1875. To the north, a row of *Liriodendron tulipifera* borders the Arboretum. The oldest tree on the island can also be found here: another tulip tree, planted by Nicolas Esterházy in 1830. In the south, a historic avenue lined with chestnut trees, sycamores, oak trees and walnut trees runs from the equally historic reservoir to the so-called Lime-Tree "Viktoria". In the first months of a flower season, this section turns into a true spring avenue when thousands of spring geophytes, especially daffodils and tulips, flower on either side of the lane.

Looking up from the bottom of the trunk to the crown of a *Sequoiadendron giganteum*: In 1864, Frederick I planted the first of these conifers native to California in today's Arboretum. More were added over the years, so that today over 60 of these giant trees can be found on Mainau, turning the island into a hotspot for this species in Germany.

Mainau is famous for its lawn areas where thousands of tulips appear in spring. This unusual display was created by necessity: In the early 1950s, an early onset of winter in November with low temperatures let the ground freeze. Only on the southern slope, the ground was covered by leaves instead of snow, so all left over bulbs were planted here.

New tulips are added to every so-called tulip field after a maximum of two years. With the bulbs slumbering in the ground, ever changing flower impressions appear as they awake.

Following page: From the flower bed by the Gardener's Tower with crown imperials, tulips and daffodils, the eye wanders past a group of *Cedrus atlantica* and a vast clearing – the so-called Long View – to the Arboretum.

In the Arboretum, clearings and views alternate with tree groups and solitary trees. Grand Duke Frederick I laid the dendrological foundation, but it was Count Lennart Bernadotte who gave the tree collection the appearance it has today. After the death of Frederick I in 1907, the park fell into a long slumber and due to the dense planting and lack of care it had soon become overgrown. By systematically removing some less valuable trees and shrubs, Count Bernadotte succeeded in turning the botanical chaos into a park.

In 1870, at least one specimen of *Catalpa bignonioides* native to Southeast North America was planted which was given enough space to develop in the Mainau Arboretum. Through natural layering, the splendid summer bloomer has spread out considerably and with its twigs and branches it has an enchanted appearance. It is often compared to a somewhat lazy person. After all, *Catalpa* is late in leafing out and then has an early leaf drop.

Cladrastis lutea with its glowing yellow autumn colour stands out particularly well in front of the backdrop of a dark yew tree and a false cypress. The tree which is native to Eastern North America and belongs to the papilionaceous plants produces pendulous clusters of striking white flowers in late spring.

Betula albosinensis is a tree with a particularly interesting bark structure. The bark of this tree native to western China peels off in fine sheets and its distinctive feature is its copper coloured trunk.

AT THE HARBOUR

The historic harbour of Mainau Island is among the most frequented in Lake Constance. About one third of all visitors enter the park here after reaching the island with a ship of the White Fleet. They are greeted by impressive giant sequoias going back to the time of the Grand Dukes.

From 2012 to 2014, the entire harbour entrance area was reshaped. Essentially, a new restaurant was built at the foot of the castle, standing out due to the interesting architecture of its façade of silver fir timber. To the building with the name of Comturey was added a smaller building with a beer garden and here is also located the information desk and the ticket offices. The name goes back to the Teutonic Knights who ruled here as commanders (Komture). The roof of the single-storey Comturey is used as an intensive green roof with small trees as well as areas with shrubs and bushes, spreading from the slope at the foot of the castle and over the building. As a result, the house blends more and more into the surrounding green and even when they are on the rooftop, the visitors do not realize they are on a building. When the garden was built, the existing path seamed with hydrangeas was extended beyond the building to the Shore Garden, allowing visitors to reach the upper level of the island without having to climb any steps. The horticultural highlights on the roof garden are, on the one hand, grey leaved trees, shrubs and bushes and, on the other hand, the presentation of the hydrangea assortment which was considerably increased in the course of the reorganization on the island.

In the course of the reshaping of the harbour, a new beer garden was built in 2014. The criteria for the trees that were to be planted on the roof were that they shall provide shade but not obstruct the view, so a rather untypical tree for beer gardens was chosen: *Albizia julibrissin* .Its leaves sprout rather late, but it has very fine flowers in summer.

In order to allow natural light to enter into the back of the building and give the historic tower more space, the architect of the Comturey, Matthias Hein, created four fully glazed, unroofed lightwells. In these garden areas, the garden designers created small green oases allowing more sensitive plants like *Fatsia japonica* to grow here due to the special climate.

At the bottom of the slope in front of the castle, the roof garden of the new building stretches out with the historic Comturey tower in the centre. The large main path which opens up onto a square in front of the tower, crosses a garden characterized by species-rich plantations of trees and shrubs with umbrella-shaped trees providing shade. For areas with a lot of heat reflection, mainly grey-leaved plants were chosen.

Hydrangea serrata 'Avelroz' is one of the more recent breedings of this hydrangea group. The inflorescence of this shrub native to Japan comes across as very natural, for this reason the plant can easily be combined with other trees, shrubs and bushes. It grows well in a sunny place or semi-shade where the humus rich, sufficiently moist but still permeable soil has a moderate pH value.

At the western end of the roof garden, where the main path gradually ascends to the more elevated areas of the island, the selected plants are predominantly classic *Hydrangea macrophylla* with their ball-shaped flowers. Over 50 different cultivars from this group alone, such as 'Bodensee' or 'Bloomstar' from the series "Endless Summer", were arranged with companion plants to create harmonious colour compositions.

IN THE SHORE GARDEN

The so-called Shore Garden stretches at the bottom of the island's eastern slope between Comturey and the Italian Style Floral and Water Cascade. This part of the park is, on the one hand, structured by different trees on the lakeshore such as *Fagus sylvatica, Buxus sempervirens* and *Taxus baccata* and, on the other hand, by conifers planted on the eastern slope such as *Chamaecyparis lawsoniana* and *Thuja plicata*, some of them very impressive. In between grow some interesting solitary trees like a *Juglans nigra* or a *Metasequoia glyptostroboides*. The latter is a present of Kew Gardens, botanical garden in London, and was planted here in 1952.

In the Shore Garden are also generous plantings of different species and varieties of *Rhododendron*, rather untypical for this region. *Camellia japonica* and its cultivars thrive well in the specially created planting beds, there is no need for elaborate protection measures during winter and the plants bloom regularly.

A narrow belt of flowers winds through the Shore Garden, displaying mainly late blooming tulips in spring – and in summer, old shrub fuchsias measuring almost three metres and other potted treasures from the nursery are presented in combination with summer flowers. On an inclined surface of the eastern slope, a carpet of flowers was created in 1975, representing Lake Constance and the most important cities on its shores. At the north end, flanked by mighty lime trees which are probably over 200 years old, is the Floral Clock. This special flower bed was arranged according to the flower clock created by Carl von Linné. The flowers are planted in a way which allows to read the time as the individual blossoms open and close.

In the long flower bed in the Shore Garden, the lily flowered tulip 'Yume no Murasaki' and the fringed variety 'Fancy Frills' are blooming surrounded by the white tulips 'Spring Green' and 'Purissima', the white daffodils 'Stainless' and 'Cheerfulness' and the blue *Anemone coronaria* 'Mr Fokker'.

In May, the Rhododendron bushes on the Flower Island are in full bloom, precisely when the floral calendar indicates the transition from spring to summer flowers. When the first Rhododendron bushes were planted on Mainau in 1971, the gardeners had to go to great lengths at the time to prepare the soil for these shrubs

growing in a low soil pH. And it was worth the effort: The individual bushes developed into harmoniously entangled groups of shrubs and in the blooming period, their colourful flowers stand out between the green of the conifers growing on the slope.

Fuchsia, like this variety 'Beacon Rosa', are a classic among the potted plants and on Mainau they have a long tradition.

Over time, the flower bed design on Mainau Island has changed: perennial plants, for example, are also used to enrich traditional flower beds today. This also includes the hybrids of *Echinacea*.

Canna indica particularly represents tropical exoticism and is a firm component of the summer flower assortment on Mainau. In the Shore Garden flower belt, these ornamental plants stand out exceptionally well. The high variety 'Marabout' displays orange-red flowers.

THE ITALIAN STYLE FLORAL AND WATER CASCADE

The Italian Style Floral and Water Cascade on the south side of Mainau Island is one of the most photographed locations on the Flower Island. It was created from 1981 to 1982 on the occasion of Count Lennart Bernadotte's 50th company anniversary according to plans by Josef Raff. The Italian Style Floral and Water Cascade starts at the southeastern end of the Mediterranean Terraces. The water originates just underneath an overlook, descends many steps into the Dahlia Garden, overcoming a difference in height of around 18 metres. Like the Fountain Arena and the Mediterranean Terraces, the Italian Style Floral and Water Cascade is made out of Calanca gneiss whose anthracite background contrasts beautifully with the multicoloured flower display. The flower display by the Cascade also changes twice a year: In spring, tulips, daffodils and other spring bloomers brighten up the Cascade with their colourful flowers and from May, subtropical flowering shrubs like pomegranate, *plumbago auriculata* and bougainvillea spread a southern flair. Mediterranean cypresses on both sides highlight the architectural effect of the rectilinear watercourse and a large number of lavender bushes on the slope emphasizes the long-distance effect of the rotating bedding plants.

Originally, the garden as it had been planned by Josef Raff ended with the lower basin by the path leading from the dahlias into the Shore Garden. However, a few years ago, the watercourse was extended from the lower basin to the lakeshore for an art project by the German artist Stefan Szczesny. It implied lining the bottom of the entire water channel, from the source to where it flows into the lake, including the basins, with a gold-coloured glass mosaic. Additionally, in three water basins, black tesserae were used to create nude drawings typical of the artist.

In spring, framed by Mediterranean cypresses, the flower beds of the Italian Style Floral and Water Cascade, created in the style of the Renaissance garden cascades, spread out to the lake like a carpet. *Viola cornuta* is a low-maintenance and sturdy as well as a vividly coloured plant and is very suited to planting under tulips.

The water to irrigate the plants is pumped out of Lake Constance with a pump station at the northern shore and distributed over the island by means of a water supply network. The gardeners collect the water from hydrants and water the plants with swing sprinklers or hoses. Sometimes the sunlight, plantings and the water spray create visually fascinating scenes.

If on Mainau the dahlia is called the Queen of Autumn, then he is the King of the third season: *Liquidambar styraciflua*. Native to North and Central America, the tree wears its magnificent autumn dress and towers over the Italian Style Floral and Water Cascade in luminous colours. At its feet stretches a peony garden, fascinating the visitors in May when varieties such as *Paeonia rockii* are in bloom.

Following page: Morning impression in midsummer at the Italian Style Floral and Water Cascade: The pale violet crowns of *Lantana montevidensis*, specimen plants grown into standard form, float over low-growing plants in matching colours, among them the dahlia variety 'Melody Fanfare', *Pennisetum setaceum* 'Rubrum', *Plectranthus scutellarioides* and *Salvia farinacea*.

IN THE DAHLIA GARDEN

At the suggestion of the longtime president of the German Dahlia, Fuchsia and Gladiolus Society, Hans Guenther Moes, the first dahlia exhibition took place on Mainau in 1952. Over 10,000 plants in approx. 200 varieties were on display in the Shore Garden. Thus the foundation was laid for an annual flower exhibition, very popular with the visitors of the island.

But over the years, the Shore Garden became too shady for the sun-loving dahlias, so they moved to the specially designed new Dahlia Garden at the bottom of the Mediterranean Terraces in 1972. Finally, in autumn 2000, the rectangular fields to the south of the main path were turned into two curved belts.

From the end of August to the beginning of October, this is the setting for the annual dahlia exhibition with more than 250 varieties. Every year in April, the Mainau nursery receives rooted cuttings from German and other European cultivators, which are then raised to obtain seedlings for the exhibition. These are planted in the Dahlia Garden in the second half of June. This means that the dahlias will not be in full bloom until the end of August, but then another floral firework will erupt at the end of the flower season.

Underneath *Cedrus libani*, the dahlia varieties 'Jacqueline' and 'Naturarena Bergisches Land' among others are patiently waiting for visitors. Every year, the visitors vote for their favourite dahlia to be crowned the "Mainau Queen of Dahlias". The plants in the foreground are, among others, *Salvia farinacea* and *Ocimum basilicum*.

When the dahlias are in full bloom at the end of the flower season on Mainau Island, the deciduous trees and shrubs on the southern slopes also dress in their colourful autumn attire: the golden yellow foliage of *Cercidiphyllum japonicum*, the yellow-green leaves of *Gymnocladus dioicus* and the red colour of *Cotinus coggygria* 'Royal Purple' complete the dahlia fireworks.

The variety 'Alexander von Humboldt' from the group of decorative dahlias belongs to the standard repertoire of the Mainau dahlia exhibition. After all, it was already crowned Mainau Queen of Dahlias in 1964.

The relatively late planting date of dahlias brings along an interesting design aspect for the gardeners: In spring, the vast flower beds in the Dahlia Garden can be used to display spring flowers that would not blossom in a classic spring flower bed on Mainau Island due to their late flowering date. Late tulips, allium and the Iceland Poppy complement leopard's bane, sweet rocket and honesties, creating attractive plant combinations.

In the Dahlia Garden, a melange of simple late tulips such as 'Queen of Night' in black red, 'Kingsblood' in dark red, 'Portofino' with cream and red flames and 'Princess Margaret Rose' in yellow with red edges open up the flower season together with *Myosotis sylvatica* 'Snowsylva' and *Papaver nudicaule* 'Pulcinella Improved Red'.

Early in the morning, the dahlias glow as the southern shore of Lake Constance is already exposed to the sun. In order to preserve this delightful impression until October, the gardeners attach the tubers to poles and regularly remove the dead heads.

89

PERENNIAL GARDEN AND SHRUB ROSES

To the west of the Dahlia Garden is the Walk of Wild and Shrub Roses. The collection of different rose varieties, selections and cultivars was created by Josef Raff, garden director at the time, from 1969 to 1979. The uncontested rose expert wanted to observe the growth of shrub roses in practical tests with different pruning measures and planted initially three specimens of each variety. The shrubs that were not pruned at all became very big, giving the collection its distinctive character to this day.

Over the years, the collection was regularly rejuvenated and complemented. The maintenance of the garden and the choice of new plants have been adapted to observe new stipulations regarding plant protection. In its commercial activity, Mainau GmbH particularly focuses on sustainability. For this reason, only biological plant protection agents and plant strengtheners are used in the Walk of Wild and Shrub Roses today. Beautifying the rose display is a backdrop of woody plants such as a group of *Taxodium distichum* and several *Sequoia sempervirens* which the island's gardeners bought from the sequoia farm in Kaldenkirchen in 1956. At the feet of these sizeable conifers, the new Perennial Garden of Mainau spreads since 2010. This rather new part of the garden was created at the initiative of the German Perennial Plant Association ("Gesellschaft der Staudenfreunde") and based on a concept by Prof. Dr. Stefan Koerner of the University of Kassel. The aim was to create a garden which would be highly appealing all year round, especially for flower lovers, by choosing the right plants. Very fittingly, the underlying motto of the project was "Fireworks in slow motion".

Despite the fact that it is on a slope, the new Perennial Garden with almost 4,000 m² is wheelchair accessible due to two stepless main paths running parallel to the Walk of Wild and Shrub Roses. Subordinate paths and paths for the gardeners structure the individual flower beds. The clipped hedges *Taxus media* 'Hicksii' and *Fagus sylvatica* create space and provide a framework for the planting beds.

The orange-red flowering *Crocosmia paniculata*, a perennial plant native to South Africa, creates an exotic flair with its striking blossoms. It is one of the hardiest specimens of this species in the Perennial Garden.

In late summer, *Rudbeckia fulgida* var. *deamii* is one of the dominating plants in the Perennial Garden, here in combination with *Calamagrostis × acutiflora* 'Waldenbuch'.

Many varieties of *Hemerocallis*, among them novelties such as the bright orange flowering 'Blazing Lamp Sticks', were planted in the Perennial Garden. The garden designers consider the low-maintenance perennials to be the new classics in perennial use.

Following page: In May, many different varieties of peony flower in the Perennial Garden. Here *Paeonia lactiflora* 'A. Delatour' is accompanied by *Iris sibirica* 'Sibtosa Princess' and *Euphorbia seguieriana* ssp. *niciciana* among others.

In the Walk of Wild and Shrub Roses, rambler roses were also regularly planted, developing over the years into stately shrubs without trellises, such as the variety 'Bantry Bay' on the left in the picture. To its right, the shrub rose 'Heinrich Conrad Söth' is in full bloom and in the background can be seen the classic 'Westerland' and *Rosa alba* 'Semiplena'.

Rosa roxburghii is one of the collection's botanical treasures. The shrub with its delicate blossoms and the exfoliating bark, typical of this species, is also suitable as a decorative winter shrub.

BY THE POND

The shore area between the Dahlia Garden and the Farm has deliberately been designed in a more natural manner. Groups of native trees and shrubs like *Alnus glutinosa, Salix alba, Populus nigra* and *Fraxinus excelsior* are side by side with exotic specimens such as *Pterocarya fraxinifolia* and *Taxodium distichum.* Walking along the seawall, the visitor hardly notices that this area was only fortified in the years 1861 to 1897 with the construction of the seawall by Grand Duke Frederick I. The pond, which was constructed in 1962 and includes a fountain, is also on the lakeside path. In front of the backdrop of trees and shrubs grow native as well as other perennials such as *Lythrum salicaria, Iris pseudacorus* and *Gunnera manicata.* A bit further to the east, a near to nature perennial plantation was created here in spring 2016. For this plantation, perennials, trees and shrubs with large leaves were chosen in order to emphasize the enchanted feel of the place.

In the park on Mainau Island there is a lot to discover, especially for children. Not far from the pond on the island's southern shore, this forest spirit created by Franz Jaeger hides behind bamboo canes, the old stump of a sequoia tree is its home.

In 2013, a new Planted Statue came into existence by the side of the Walk of Wild and Shrub Roses. The artwork with the title “Together in the Garden” was designed and built by the Dutch artist Sjer Jacobs and differs from the other flower figures in its set-up and also in the spring and summer flowers used. In the background, the bald cypresses are losing their last needles, bearing autumn colours.

The pond with its naturalistic plantings is a place of peace and quiet and natural calmness. This is reflected by the grass carps leisurely swimming in the pond. The heron searching for food in the pond is a carved wood artwork created by an employee of Mainau Island, Markus Pfeiffer. The pond is fed by the fountain with water from the garden network and has an overflow into Lake Constance.

LAUEN AREA

Visitors coming to Mainau Island from the mainland are greeted by the flower figure Blumi, the Flower Island's mascot, at the southwestern tip of the island, the so-called "Lauen". The Mainau-Kinderland is just a step away, an adventure world for the smallest visitors, including the playgrounds "Dwarf Village", "Blumi's Coast Adventure" and "Water Adventure" as well as a miniature railway, the farm and the petting zoo.

The Flower Animals in the shape of a peacock and a duck family are a permanent feature of the park and for many visitors the symbol of Mainau Island. The duck family swims peacefully in a pond and flowers also spread across the water's surface. Inside the figures is a construction of steel round bars bent into shape. The construction is tightly filled with potting soil and covered with mesh wire. That way the soil is firmly in place and flowers can be planted – horned violets or daisies in spring and wax begonias in summer.

In the north, the vast "Lauen" is bordered by the *Metasequoia* Avenue. Due to its age and dimension, it is considered one of the island's dendrological treasures. Planted in 1959, their bizarre tree base particularly impresses today's visitors.

The playground "Dwarf Village" is reserved for the island's youngest visitors. It is framed by high-stem *Prunus serrulata* 'Kanzan' and sitting under their crowns is a crooked fishing hut, the "Blumi's Coast Adventure" playhouse. This play area symbolizes the Lake Constance landscape and is meant to sharpen the visitors' senses by means of elements like the thread-maze ("Fadenlabyrinth").

The duck family is popular with old and young visitors alike and a much-loved place to take photographs. Because of their compact growth and profusion of flowers, horned violets are particularly suitable for spring planting, followed by wax begonias in summer. The water surface of the pond is covered in blue horned violets and blue ageratum.

Since 2003, in the section "Im Lauen" the visitors are greeted by the floral figure "Blumi". In front of the figure, the flower heads of *Fritillaria persica* 'Ivory Bells' stick out between a group of Single Early Tulips 'Red Revival' and Triumph Tulips 'Black Jack' and 'Christmas Yellow'.

On the meadow between the playground and Lake Constance, the fresh green leaves of *Salix alba* 'Tristis' are glowing in the sunlight. At its feet, the daffodil variety 'Accent' and a number of late tulip varieties are among those providing a play of colour in the tall grass.

As the Sycamore Avenue from the bridge to the castle had become too big and the road too narrow, a new, wider avenue lined with Dawn Redwoods was laid out from 1959: The *Metasequoia* Avenue was born. The trees were grown from cuttings from the *Metasequoia* in the Shore Garden which rapidly turned into impressive specimens. Today, the Avenue is one of

the dendrological highlights, inspiring in every season: in spring, when countless geophytes such as *Crocus tommasinianus* are blooming or in summer, when the trees provide dappled shade.

FROM PLATANENWEG 5 TO THE KITCHEN GARDEN

At the northwestern end of the *Metasequoia* Avenue is located Platanenweg 5, a section of the park with a typical cottage garden design. In fact, the entire park on Mainau Island is a place of inspiration and an endless source of ideas for garden and plant lovers. In Platanenweg 5, visitors obtain even more gardening information: In a garden house, the gardeners provide tips every day in the flower season, answering all questions about plants and gardens. in the garden itself, new plant varieties, combinations and garden trends are presented on a regular basis. Besides the traditional perennial plantings and changing flower beds, the plant lover will also find suggestions here for floral compositions for boxes and pots on their balcony or terrace or further information on specific gardening topic during temporary exhibitions and events.

On the other side of the large main path is situated the Garden for Everyone, built in 1989 and inaugurated by H. M. Queen Silvia of Sweden. The garden with its somewhat unusual design was initiated by Countess Sonja Bernadotte for people with a disability. Here, the disabled and their relatives obtain tips and information on how to enjoy gardening from a wheelchair or which plants are suitable for a blind person's garden.

Next to the Garden for Everyone is the Café "Vergissmeinnicht", run by young people with special needs and preparing them for working life. Behind it is the Kitchen Garden, until 2013 this area was used for horticultural production, now vegetables, fruit and flowers are grown and harvested in this instructive garden. According to old maps, the Teutonic Knights already had gardens where they grew fruit and vegetables to feed the island's population.

Even if the vegetable island of Reichenau is only a few kilometres away – the Mainau park also has vegetables, often used to structure the flower beds. Here, chard is grown with red cabbage in the Kitchen Garden. In the background *Tagetes tenuifolia* is blooming and the leaves of *Atriplex hortensis* 'Rubra' are a lovely shining red.

A map of Mainau Island in the 18th century after the castle was built shows geometrically designed gardens in the area of today's Arboretum and it is most likely that they not only had a representative function, but also provided food for the commandery of Mainau. Inspired by these gardens, the new Kitchen Garden was built next to the Café "Vergissmeinnicht". Vegetables, summer flowers and fruit are grown and harvested here.

For Countess Sonja Bernadotte, helping young people with special needs was very close to her heart. She did not want to exclude disabled persons from gardening. For this reason she had the idea for the Garden for Everyone. Apart from a sensory garden for the blind, the inclusive garden also comprises areas with raised beds where *kniphofia* and other flowers bloom in summer. The raised beds allow easy wheelchair access to the plants.

The concept of Platanenweg 5 is that of a traditional cottage garden with perennials and flower beds, a terrace with pergola and a garden house. Over the course of the year, presentations of new products for home gardens and exhibitions on different plant topics regularly take place in the entire garden. The deck chairs are part of the information and not only display items. They invite visitors to relax and enjoy.

The trend towards using timber stacks to create walls as in the house garden was also adopted in Platanenweg 5. The planters in front of the garden house are an inspiration for visitors with boxes and pots at home on a balcony or terrace. Additionally, visitors can obtain information from gardeners who are in the garden house every day during the flower season to give expert advice on all kinds of plants.

BUTTERFLY HOUSE AND INSECT GARDEN

Inspired by a trip around the world, Countess Sonja Bernadotte and Count Lennart Bernadotte built a tropical Butterfly House which opened in 1996. In a specially built traditional greenhouse, Mainau Island gardeners created a tropical landscape of almost 900 m^2 with unusual plants, ponds and waterfalls. But the stars of the greenhouse are the tropical butterflies. The creatures of beauty reach the island as pupae, coming from butterfly farms in Great Britain, Costa Rica and Malaysia where they are bred for many butterfly houses all over the world. Over the course of one year, about 120 butterfly species can be watched and enjoyed in the Butterfly House. Depending on the species and size, a butterfly lives between one or two weeks and three months. Due to their many years of experience, the gardeners now successfully breed some butterfly species directly in the Butterfly House – especially because the island's nursery nearby allows the cultivation of plants the butterflies particularly enjoy as food and nectar plants. These plants include *Asclepias curassavica* as well as *Cestrum* spec. or *Senna corymbosa*.

The open-air equivalent to the Butterfly House is the Insect Garden, created in spring 2012 right next to Platanenweg 5. This nature-oriented park section is designed to provide information about native insects, especially honeybees, wild bees and butterflies. Different works of art dedicated to the topic are displayed in herb and flower rich meadows. Inspecting the man-size cavity in the trunk of an old poplar, a beehive conveying the energy, warmth and scent of the bee colony, is an extraordinary experience.

From the corn poppies in the Insect Garden, the visitor's gaze wanders over the vineyard to the Swede's Tower. The vineyard which was rebuilt in 2006 on the original site is planted with the two red grape varieties 'Cabernet Cortis' and 'Monarch' as well as the white varieties 'Solaris' and 'Johanniter'. These varieties are very resistant to downy mildew, thus allowing ecological viticulture.

The Insect Garden was created in 2012 with the aim of informing the island's visitors about the life of our insects and the threats they encounter, especially wild and honey bees. Another aspect is the importance of animals for the pollination of our crops. For this reason the garden, which develops into a flower-rich meadow over the years, was built right next to an existing orchard.

Two walls made out of old natural and artificial stone as well as pieces of wood are particularly eye-catching. Their niches, gaps and holes provide shelter and a nesting aid for wild bees and other insects, but they are also popular with lizards. In front of the attractive construction, the flowers of *Echium vulgare* are glowing in a fresh blue colour.

In the Butterfly House, the climate allows tropical plants to thrive all year round. By the large pond with the waterfall grow *Monstera deliciosa* and *M. adansonii*, *Coffea arabica* and *Carica papaya*, the latter regularly bearing its tasty orange fruits.

Dryas iulia, also known as The Flame, is a diurnal butterfly from the Nymphalidae family native to Central America. Its caterpillar feeds on leaves of different passion vines.

CONTACT

Mainau GmbH
78465 Insel Mainau
Telefon 07531/303-0
Telefax 07531/303-248
E-Mail: info@mainau.de
www.mainau.de

OPENING TIMES

Mainau Island is open all year round from sunrise until sunset.

THANKS

The author thanks Nina Busse, Lisa Boehnke, Jutta Hardt and Florian Heitzmann for their support.

SOURCE OF PHOTOGRAPHIC MATERIAL

All photographs by Peter Allgaier
Map of the island Page 8: Mainau GmbH

LEGAL DETAILS

The suggestions and information provided in this book were compiled and verified by the author with the greatest care. However, it is not possible to guarantee all information is correct. The author and the publisher assume no liability for damages or accidents. When implementing the suggestions contained in this book, please use your personal judgement.
The publisher Eugen Ulmer is not responsible for the contents of the websites mentioned in this book.

Bibliographic information of the German National Library
The German National Library registers this publication in the German National Bibliography; detailed bibliographic data is available on the internet via http://dnb.d-nb.de.

Wollgrasweg 41, 70599 Stuttgart (Hohenheim)
E-mail: info@ulmer.de
Internet: www.ulmer-verlag.de
Translation from German into English: Miriam Schmid/ ade team übersetzungen
Proofreading: Christine Condé, Helen Haas
Production: Ulla Stammel
Jacket design: Ulmer Verlag
Composition: BUCHFLINK Rüdiger Wagner, Nördlingen
Printing and binding: aprinta Druck, Firmengruppe APPL, Wemding
Printed in Germany

ISBN 978-3-8186-0041-9